BUILDING INFORMATION MODELLING (BIM)

WHY BIM FAIL

THE HUMAN ERRORS BEHIND DIGITAL TRANSFORMATION DISASTERS

ABDUL RAFIQ GANASAN BIN ABDULLAH

notionpress.com

INDIA · SINGAPORE · MALAYSIA

ISBN 979-8-90342-634-8

WHY BIM FAILS TO DELIVER LOD 500

A Condensed Guide to

Systemic Failure Across All Players

Table of Contents

Preface

Writing a book is never a solitary endeavour. Though one name appears on the cover, the pages within carry the invisible fingerprints of many—those who supported, encouraged, corrected and guided along the way.

It is with genuine gratitude that I offer my thanks to the people who made this book possible.

To My Family

First and foremost, to my beloved wife, Hasnah Bte Ros—your patience through the long hours, the late nights, and the endless conversations about BIM has been my foundation. You read the proofs, offered corrections, and never once complained when the book demanded more of my time. This book is as much yours as it is mine.

To my daughters and son, thank you for understanding when father was "working on the book" yet again. Your love and encouragement carried me through. I hope this work makes you proud.

To My Friends

To Faizal—your friendship across these many years has been a gift. Thank you for reading, for questioning, for pushing me to be clearer and more precise. A friend who challenges is a friend indeed.

To Guna—your perspective and suggestions improved this book in countless ways. Thank you for the conversations, the critiques, and the encouragement.

To My Lecturers and Mentors

No one reaches thirty-five years in this industry without standing on the shoulders of those who taught them. I am deeply grateful to the educators who shaped my understanding and gave me the tools to write this book.

To Ts. Aiman and Ts. Ashraff of MTTC Malaysia – your teaching opened doors I didn't know existed. You didn't just teach BIM; you taught me to think about BIM. Your influence runs through every chapter.

To Ts Aidzil Adzahar Ahmad of JKR Malaysia, your guidance has been invaluable. You showed me that BIM is not just about technology, but about process, people, and purpose. Your insights gave me the confidence to write this book, and your standards pushed me to make it better.

A Special Obligation

I would like to record here my special obligation to:

Ts Aidzil Adzahar Ahmad
Ts Aiman
Ts Ashraff

Their teaching and guidance gave me the strength to write this book. In a career filled with lessons, theirs stand out as among the most important. They taught me not just what BIM is, but why it matters. Not just how to implement it, but how to think about it. Not just the technical, but the human.

This book exists because of them. Whatever value readers find in these pages, some portion of that credit belongs to these three men.

To the Reader

Finally, thank you for picking up this book. Whether you are a client trying to understand what LOD 500 really means, a consultant frustrated by failed handovers, a contractor tired of rework, a BIM Manager fighting for authority, or an FM team member tired of receiving useless models—this book is for you.

The industry can do better. We can do better. This book is my contribution to that effort.

May it serve you well.

Abdul Rafiq Ganasan bin Abdullah
Johor, Malaysia
2026

Part I: The Promise

Chapter 1:

What LOD 500 Should Be

1.1 Defining the Destination

Before we understand why BIM fails to deliver LOD 500, we must understand what LOD 500 actually means. Level of Development 500 represents the culmination of the BIM process—the model that represents the building as it was actually constructed. Not as designed, not as tendered, as-built.

LOD 500 rests on three essential pillars:

Pillar 1: Geometric Accuracy

Every element in the model must reflect the actual installed location, size, and shape. If a pipe was moved 300mm to avoid a beam, the model shows that the pipe is 300mm offset. The geometry is not "design intent"—it is "as-built reality."

Pillar 2: Informational Completeness

Every element carries data. The pump is not just a 3D shape—it contains manufacturer, model number, serial number, installation date, warranty expiry, maintenance schedule, and links to manuals. The wall contains fire rating, acoustic properties, material composition, and installation contractor.

Pillar 3: Operational Usability

The model must be usable by facilities management. This means compatibility with their systems (CAFM/CMMS), organisation according to their asset hierarchy, and structure for their workflows. A model that cannot be used for operations has failed to meet LOD 500.

1.2 Common Misconceptions

Misconception 1: "LOD 500 is just a more detailed LOD 400"
False. LOD 400 is for construction. LOD 500 is for operations. They serve different purposes and require different information.

Misconception 2: "LOD 500 means everything is modelled in high detail"
False. LOD 500 requires appropriate detail for operations. The principle is "right information, right place, right time"—not "everything modelled to death."

Misconception 3: "LOD 500 happens at the end"
False. LOD 500 must be built continuously throughout construction. Waiting until the end guarantees failure because information has already been lost.

1.3 Level of Information Need

Under ISO 19650, we no longer assume that LOD 500 means the same thing on every project. The Level of Information Need (LOIN) framework asks:

- **Why:** For what purpose? (Operations? Maintenance? Renovation?)
- **When:** At what stage? (Handover? During warranty?)
- **Who:** Who provides? Who receives? Who verifies?
- **What:** What specific information? (Geometry? Data? Documents?)
- **How:** How detailed? How accurate? What format?

A project that fails to define its LOIN cannot deliver LOD 500 because no one knows what LOD 500 means for that project.

Chapter 2:

The Value Proposition

2.1 Why Should Anyone Care?

For Owners:

- Reduced operating costs through accurate asset data.
- Faster response to issues with instant asset location.
- Higher asset valuation with complete digital records.
- Regulatory compliance, demonstrable through as-built conditions.

For Contractors:

- Reduced rework by catching errors before they're built.
- Fewer disputes with documented as-built conditions.
- Competitive advantage through reliable delivery
- Risk reduction with accurate records

For Facility Managers:

- Day one readiness from the moment they take possession.
- Efficient operations with assets found in seconds.
- Career satisfaction, doing the job they were hired to do.

The Cost of Failure:

- Manual re-creation of as-built information
- Site surveys to verify conditions
- Extended handover periods
- Delayed facility commissioning
- Inefficient operations over the building's lifetime
- Premature equipment replacement
- Lost opportunity for optimisation

Part II: The Players and Their Failures

Chapter 3:

The Client – "I Asked for LOD 500, I Received Chaos"

3.1 How the Client Fails

Failure 1: Requiring Without Understanding

Clients read about BIM, hear "LOD 500 is the gold standard," and include it in tender documents. But no one asks: What does LOD 500 mean for this specific project? What assets need data? What format? What software will FM use? The contractor delivers a model. It has geometry. It has some data. But it doesn't work for FM.

Failure 2: Not Allocating Budget

BIM costs money. LOD 500 costs a significant amount of money. Someone must pay for modelling, data collection, verification, software, and training. Clients who don't allocate a budget are asking for something they haven't paid for.

Failure 3: Late Involvement of FM

The FM team knows what data they need. They are seldom involved until the building is complete. By then, the data structure is set, software choices made, and information collection finished. The FM team receives a model designed by people who have never operated a building.

Failure 4: No Internal Capability

Even with a perfect model, many clients cannot use it. They have no software to open it, no staff trained to navigate it, no IT infrastructure to host it, and no processes to maintain it.

Failure 5: Vague Contract Language

"I want LOD 500" is not a requirement. Without specificity—which assets, what data, what format, when, how verified—these words mean nothing.

3.2 What Good Clients Do

- Educate themselves on BIM basics
- Involve FM from the design stage
- Budget appropriately for all BIM activities
- Write specific, enforceable requirements
- Verify throughout construction, not just at handover.
- Build internal capability before handover
- Appoint a BIM representative

Chapter 4:

The Consultant – "Design in Fantasy, Deliver in Denial"

4.1 How the Consultant Fails

Failure 1: Designing for Tender, Not for Construction

Design models serve to communicate intent and secure approvals. They lack detail for construction coordination. They contain assumptions that will change. When handed to contractors, problems begin.

Failure 2: Abandoning the Model After Design

Many consultants believe responsibility ends when design is complete. When construction reveals issues, the model isn't updated. When substitutions occur, it isn't revised. The design model becomes increasingly irrelevant.

Failure 3: Modelling Without Construction Knowledge

Talented modellers create beautiful geometry but have never been on site. They don't know installation space requirements, trade sequences, or access needs. Models are geometrically correct but practically impossible.

Failure 4: No Data Strategy

Consultants are comfortable with geometry, not data. Their models contain generic placeholders ("TBC"), specification references that become obsolete, and no structure for as-built information.

Failure 5: Over-Modelling and Under-Informing

Consultants model every bolt, but don't include basic data like fire ratings. Visual impact wins awards. Information doesn't. Information gets neglected.

4.2 What Good Consultants Do

- Design for buildability with construction knowledge.
- Stay involved through construction.
- Build information structure with data fields ready.
- Collaborate with FM to understand operational needs.
- Model appropriately for each purpose.
- Ensure models are usable by contractors.

Chapter 5:

The Main Contractor – "The Middleman Who Drops the Ball"

5.1 How the Contractor Fails

Failure 1: Treating BIM as a Tender-Winning Tool

BIM is used for impressive bid visuals. Once the contract is signed, BIM becomes overhead to minimise. The delivery team has different priorities. Promises become obligations to avoid.

Failure 2: No Integration Between Site and Office

BIM team sits in the office. Site team works on-site. Between them lies a communication gap that swallows information. Changes happen on-site. BIM team doesn't know. Model remains unchanged.

Failure 3: No Contractual Flow-Down to Subcontractors

Main contractors sign contracts requiring BIM. Then they subcontract work without following the requirements. Subcontractor response: "You didn't ask for it, you didn't pay for it, I'm not doing it."

Failure 4: Treating LOD 500 as an End-of-Project Activity

Many contractors plan to "clean up the model" at the end. This is impossible. Thousands of changes occurred. No one recorded them. Subcontractors have left. Memory has faded. What arrives is a design model with "as-built" stamped on it.

Failure 5: No Verification Process

Contractors rarely verify that the model matches reality. They assume that if designed that way, it must have been built that way. Construction doesn't work like that. Without verification, the model drifts from the truth.

Failure 6: Focus on Geometry, Not Data

Contractors ensure pipes are in the right place, but they ignore data. The pump lacks a manufacturer, model number, serial number, and installation date. Without data, the model is useless for FM.

5.2 What Good Contractors Do

- Put BIM staff on-site as part of the construction team.
- Include BIM requirements in every subcontract
- Pay for BIM as part of subcontract prices

- Verify model against the site continuously
- Give BIM Managers authority to enforce requirements.
- Collect data during installation, not after
- Start handover planning at project start

Chapter 6:

The Subcontractor – "The Silent Killer of Information"

6.1 How the Subcontractor Fails

Failure 1: No BIM Requirements in Subcontract

The foundational failure. Subcontract says nothing about BIM. No modelling. No data. No as-built information. The subcontractor is not required to participate.

Failure 2: No BIM Capability

Even when required, many subcontractors lack capability. No BIM staff, no software, no training. They work the way they've always worked.

Failure 3: No Access to the Model

Even willing subcontractors cannot participate. No logins, no compatible software, no training. They receive 2D printouts—the same as always.

Failure 4: No Time for Data Entry

Site work is demanding. Every hour spent on data entry is an hour not installing. Income depends on installation, not data. Data entry is deprioritised.

Failure 5: Information Trapped with Individuals

The electrician knows where cables run. The plumber knows where valves are. When they finish, they leave. They take their knowledge with them. No one extracted it.

6.2 What Good Subcontractors Do

- Invest in BIM capability as a competitive advantage.
- Include BIM costs in bids.
- Make data collection part of installation work.
- Designate someone to manage BIM coordination.
- Train the workforce on what information matters and how to record it.

6.3 The Main Contractor's Role

- Main contractors must:
- Include BIM in subcontracts with clear requirements.
- Provide access, training, and support.
- Pay for BIM as part of progress payments.
- Treat subcontractors as partners in information delivery.

Chapter 7:

The BIM Manager – "The Prophet Without Power"

7.1 How the BIM Manager Is Set Up to Fail

Responsibility Without Authority

BIM Manager is responsible for delivering LOD 500, but cannot require subcontractors to comply, stop work for information failure, or withhold payment. They can ask. They cannot require.

Reporting to the Wrong Person

BIM Managers often report to IT or estimating, not site operations. They have no connection to people who actually build. Their requests are ignored.

Understaffed and Overworked

One BIM Manager is expected to manage models, coordinate trades, train users, enforce standards, collect data, attend meetings, and support the site. They burn out. Quality suffers. They are blamed.

No Budget

BIM Managers need software, hardware, training, staff, and systems. They rarely have a budget. They make do. Quality suffers. They are blamed.

Brought In Too Late

BIM Managers join after contracts are signed, software is chosen, and processes are set. They inherit a mess and are expected to fix it. They can't. They are blamed.

No Enforcement Mechanism

When subcontractors don't provide information, the BIM Manager can send emails, escalate, or complain. They cannot withhold payment, stop work, or impose penalties.

7.2 What BIM Managers Need

- Authority to enforce requirements
- Role in payment certification
- Direct reporting to project leadership
- Adequate staffing and budget
- Involvement from project start
- Leadership backing
- Recognition as professionals

7.3 What Good BIM Managers Do Despite the Odds

- Build relationships and trust
- Educate constantly on why BIM matters
- Create systems that make compliance easy
- Communicate relentlessly
- Document everything
- Escalate appropriately. Focus on what matters most.

Chapter 8:

The FM Team – "The Forgotten Recipient"

8.1 How the FM Team Failed

Brought In Too Late

Design: no FM. Construction: no FM. Handover: "Here's your model." FM team can't use it. Too late to fix it.

No Input to Requirements

FM teams know what they need. No one asks. Requirements are written by consultants who have never operated a building. Information is collected based on what's easy, not what's useful.

Incompatible Systems

FM teams use specialised software. LOD 500 models are delivered in formats that don't work with these systems. FM team manually re-enters data, the very thing LOD 500 was supposed to prevent.

No Training

FM team receives a file with no instructions, a model with no navigation guide, and data with no explanation. No training. The model sits on a server, unused.

Overwhelming and Irrelevant Information

FM teams don't need every bolt modelled. They need asset locations, maintenance requirements, warranty information, and manufacturer contacts. Models are delivered with everything. FM teams can't find what matters.

No Update Process

Buildings change. Equipment is replaced. Spaces renovated. FM team has no software to edit, no training to modify, no process to maintain, and no budget to update. Model becomes obsolete.

8.2 What FM Teams Need

- Early involvement from the design stage
- Input to information requirements
- Compatible systems and formats
- Training before handover
- Relevant, organised information
- Ongoing support and update process

8.3 What Good Projects Do

- Involve FM from day one
- Define requirements around FM needs
- Test with FM systems during construction
- Train before handover
- Provide ongoing support
- Plan for updates
- Have FM verify before final acceptance

Part III: The Systemic Failures

Chapter 9:

The Broken Workflow (Reason 1)

"BIM Flow Not as per Project Flow"

9.1 The Parallel Universes Problem

Universe 1: The Project

Site team solves problems daily. Trades install materials. Changes happen constantly. Progress measured in concrete poured, steel erected.

Universe 2: The BIM Model

BIM team updates weekly (or monthly), with changes reflected days or weeks later. Information is entered based on memory.

These universes should be the same. They are not. The gap grows throughout the project. By handover, they are irreconcilable.

9.2 Why Workflows Diverge

- **Timing Mismatch:** Project flow is real-time, BIM flow is delayed.
- **Priority Mismatch:** Project flow prioritises physical progress. BIM flow prioritises information. Physical progress wins.
- **Incentive Mismatch:** Project flow gets paid. BIM flow is overhead. Rational players prioritise what gets paid.
- **Communication Mismatch:** Project flow communicates through site meetings. BIM flow communicates through model updates. These channels don't connect.

9.3 The Cost of Broken Workflow

- Information loss daily
- Rework when model-based decisions are wrong
- Distrust when site teams learn the model can't be trusted.
- Handover failure when the model is too wrong to be useful.

9.4 Fixing the Workflow

- Put BIM staff on-site as part of the construction team.

- Match update frequency to change frequency (daily during rapid construction).
- Make BIM part of site meetings
- Connect payment to information updates
- Use mobile technology for field updates
- Verify continuously, keeping the gap small

Chapter 10:

The Missing Redlines (Reason 2)

"Never Practice Redline Drawing System"

10.1 The Death of Redlines

Redlines are the most basic form of as-built documentation. A tradesperson marks up a drawing with what they actually installed. This system has worked for generations. On modern BIM projects, it is seldom practised.

Why Redlines Disappeared:

- "We have BIM, we don't need paper redlines" (false assumption)
- "Just fix it and move on" (site pressure)
- Loss of craft culture that valued documentation
- Unclear responsibility
- Illusion of memory ("I'll remember")

10.2 Why Redlines Matter for LOD 500

The only people who know what was built are the people who built it. Their knowledge must be captured. Redlines are the capture mechanism.

Without redlines:

- Changes are forgotten
- Locations are approximate
- Substitutions are unrecorded
- Decisions are lost
- The model becomes design intent, not as-built

10.3 Fixing the Redline System

- Include redline requirements in every subcontract.
- Provide marked-up drawings or tablets for digital markup.
- Train that markup is part of the installation.
- Build time for documentation into schedules.
- Review redlines in weekly meetings.
- BIM team updates the model based on markups.

Chapter 11:

The Digital Native (Reason 3)

"Young Guys Good at Software, Zero Site Exposure"

11.1 The Two Knowledges

- **Software Knowledge:** How to use Revit and Navisworks. How to create geometry, manage data, and run clash detection.
- **Construction Knowledge:** How buildings go together. Trade sequences. Installation space requirements. Access needs. What's possible.

LOD 500 requires both. Without construction knowledge, the model is a beautiful lie.

11.2 Why Young Modellers Lack Site Experience

- Architecture/engineering education focuses on design, not construction.

- Career paths keep young professionals in offices.
- BIM is seen as office work, site as "different."
- No time or budget for site visits.

11.3 What Site Experience Teaches

- **Constructability:** Can this be built? Not just geometrically, but practically.
- **Sequence:** What order do things happen in? The model must reflect this.
- **Tolerance:** Nothing is perfectly straight. Site experience teaches what "close enough" means.
- **Access:** Will someone need to reach this valve in five years?
- **Reality:** The difference between a line on a screen and a pipe in a ceiling.

11.4 What Happens Without Site Experience

- Unbuildable designs that look perfect in software.
- Missed coordination, that software clash detection doesn't catch.
- Wrong assumptions about how things work.
- Lost information when the site works around the model.

11.5 Bridging the Gap

- Make site visits mandatory for modellers.
- Pair young modellers with experienced site people.
- Hire for construction experience, not just software skills.
- Include construction methods in BIM training.
- Hold BIM meetings on site.
- Walk the model against the building.

Chapter 12:

The Frozen Model (Reason 4)

"Never Update the Model from Day 1"

12.1 The Freezing Process

- **Day 1:** Model accurate (for design).
- **Week 1:** First changes, not updated. Small gap appears.
- **Month 1:** Dozens of changes. Gap grows.
- **Month 3:** Hundreds of changes. Model weeks behind.
- **Month 6:** Thousands of changes. Model irrelevant.
- **Month 12:** Project complete. Model at month 3. LOD 500 impossible.

12.2 Why Models Freeze

- No assigned responsibility for updates
- No system for capturing changes
- No incentive (no payment for updating)
- No time
- No access to site information
- No consequences for the outdated model

12.3 The Cost of Frozen Models

- Lost coordination leading to rework
- Lost efficiency when the site stops using the model
- Lost information forever
- Lost trust in BIM
- Lost value at handover

12.4 Keeping the Model Alive

- Make updates mandatory, tied to payment
- Update daily during rapid construction
- Provide tools for field updates
- Review model accuracy in meetings
- Include update time in schedules
- Verify continuously

Chapter 13:

The Empty Database (Reason 5)

"Ignore to Update the Asset Data from Day 1"

13.1 Geometry vs Data

- **Geometry:** Visible, easy to check, wins awards, what BIM is known for.
- **Data:** Invisible, hard to verify, wins nothing, what BIM is actually for.

LOD 500 requires both. Geometry without data is sculpture. Data without geometry is a spreadsheet.

13.2 Why Data Is Neglected

- Contracts don't specify what data means
- No budget for data collection
- No process for who collects when
- No incentive for subcontractors

- No understanding of what data matters
- No verification

13.3 What Data Matters

For each asset, FM needs:

- Asset ID, name, location, system
- Manufacturer, model number, serial number, installation date
- Warranty expiry, maintenance schedule, and expected life
- Manuals, guides, safety data, spare parts lists
- Installation contractor, commissioning records

13.4 When Data Should Be Collected

- **At Procurement:** What was ordered, from whom, when
- **At Delivery:** What arrived, serial numbers, documentation
- **At Installation:** Where installed, by whom, when
- **At Commissioning:** Test results, adjustments
- **At Handover:** Complete verified record

Data collected at the end is too late. Information is already lost.

13.5 Fixing the Data Problem

- Define requirements early, starting with FM
- Include data requirements in every subcontract
- Collect at source during installation
- Provide templates and mobile tools
- Verify continuously with sample checks
- Tie payment to data delivery

Chapter **14**:

The Unread Contract (Reason 6)

"Never Go Through the BEP, VDC, EIR"

14.1 The Document Graveyard

BEP (BIM Execution Plan), EIR (Employer's Information Requirements), AIR (Asset Information Requirements), MIDP, VDC Plan, LOD Matrix, Responsibility Matrix—all written, approved, filed, and never opened again.

14.2 Why Documents Go Unread

- Too long (50+ pages of dense text)
- Too generic (templates from previous projects)
- Too late (written after decisions made)
- Too boring
- No enforcement
- No connection to daily work

14.3 What the Documents Actually Contain

- **BEP:** Who does what, when, what standards, what software, what deliverables
- **EIR:** What information does the client need, for what purposes, at what stages, in what format
- **LOD Matrix:** What level of development for each element at each stage
- **Responsibility Matrix:** Who provides each deliverable, who approves

These documents contain everything needed for LOD 500 success. They are ignored at the project's peril.

14.4 What Happens When Documents Are Ignored

- No clarity on requirements
- No accountability when information missing
- No standards, so models don't coordinate
- No enforcement, so requirements are optional
- No handover—no one knows what was supposed to be delivered

14.5 Making Documents Useful

- Keep them short (10 pages max)
- Make them specific to this project and team
- Write them early, before decisions

- Involve the team in development
- Reference in meetings and daily work
- Review compliance monthly
- Tie payment to compliance

14.6 The One-Page BEP

Some projects use a one-page BEP: who does what (one table), when (one timeline), what they deliver (one list), standards (one reference), who to contact (one name). Everyone reads it. Everyone understands.

Part IV: The Main Contract Context

Chapter 15:

Traditional Contracts in a Digital World

15.1 The Main Contract Context

It was written for a 2D world. They don't understand BIM. They create problems that make LOD 500 impossible.

15.2 The Gaps

- **No Model Ownership:** Who owns the model? Architect? Contractor? Client? The main contract was never mentioned.
- **No Information Requirements:** What information at handover? The contract only specifies drawings. No mention of data, models, or digital information.
- **No Payment for BIM:** Payment tied to physical progress. BIM work isn't physical, so it doesn't get paid.

- **No Model Status:** When the model conflicts with drawings, the contract says drawings prevail. This kills model authority.

- **No As-Built Certification:** Who certifies that the model represents as-built? No one assigned. No one owns it.

15.3 The Impact on LOD 500

- No requirements = no one knows what LOD 500 means
- No payment = rational contractors minimise BIM effort
- No ownership = no one is responsible for accuracy
- No authority = model can't be a definitive record
- No consequences = when LOD 500 isn't delivered, no remedy

15.4 Adapting Traditional Contracts

- **Amend Ownership Clauses:** Specify who owns the model, licensing, and handover rights.

- **Amend Information Requirements:** Define what information, what format, what standards, and when delivered.

- **Amend Payment Provisions:** Include BIM in payment schedules, tie to information delivery.

- **Amend Dispute Resolution:** Address digital evidence, model status in disputes.
- **Add BIM-Specific Clauses:** BEP requirements, coordination obligations, data standards, handover specifications.

Chapter 16:

The Liability Problem

"Who Certifies Truth?"

16.1 The Certification Problem

No one wants to sign their name to a statement that could be wrong. The building is complex. Thousands of elements. Millions of data points. One error could mean liability. So no one certifies.

Who Could Certify?

- **Contractor:** Built it, but didn't create all the data
- **Subcontractor:** Knows their work, not others'
- **Consultant:** Designed it, wasn't on site
- **BIM Manager:** Managed model, didn't build
- **Client:** Wants a model, can't certify

No one can certify the whole model. So no one does. The model is delivered "as is" with no warranty.

16.2 Why No One Will Certify

- **Risk:** If the model is wrong and someone relies on it, the certifier is liable. Risk is huge, reward zero.
- **Cost:** Certifying requires verifying every element against reality. Expensive. No one pays.
- **Knowledge:** No single person knows everything. Certification would require collective knowledge.
- **Tradition:** No one certified as-built drawings, either. BIM follows the same pattern.

16.3 The Consequences

- No accountability for accuracy
- No trust from clients
- No value—can't make decisions based on uncertified model
- No LOD 500—without verification, it isn't LOD 500

16.4 Possible Solutions

- **Collective Certification:** Each party certifies their part. Subcontractor certifies installation. Contractor certifies coordination. BIM Manager certifies management.

- **Sample Verification:** Certify that a sample has been verified. Statistical confidence is not a guarantee.
- **Progressive Certification:** Certify at milestones, building confidence progressively.
- **Insurance-Backed Certification:** Products that cover model accuracy, spreading risk.
- **Contractual Allocation:** Contract specifies who bears the risk of inaccuracy.

16.5 The Realistic Path

LOD 500 doesn't require absolute perfection. It requires reasonable accuracy based on field verification. The realistic path:

- Define acceptable accuracy
- Specify verification methods
- Document what was verified
- Report confidence levels
- Allocate risk contractually

Part V: The Way Forward

Chapter 17:

Redefining Roles for LOD 500 Success

17.1 The Client's New Role

- Define information needs clearly, starting with FM.
- Budget appropriately for all BIM activities.
- Participate in BEP development and milestone reviews.
- Enforce requirements through payment and acceptance.

17.2 The Consultant's New Role

- Design for information with data fields from the start.
- Stay involved through construction.
- Enable contractors with usable models.
- Focus on operations and long-term value.

17.3 The Main Contractor's New Role

- Integrate BIM with construction, putting BIM staff on site.

- Flow requirements down to all subcontractors.
- Capture information continuously through redlines and data collection.
- Coordinate all players toward handover.

17.4 The Subcontractor's New Role

- Build BIM capability as a competitive advantage.
- Provide information at the source during installation.
- Participate in coordination and handover.
- Own and certify your part.

17.5 The BIM Manager's New Role

- Lead the process with clear standards and workflows.
- Support the team with training and resources.
- Warn of risks and escalate appropriately.
- Deliver verified, usable information.

17.6 The FM Team's New Role

- Define what information you need and in what format.
- Participate from the design stage through handover.
- Prepare systems and staff to receive the model.
- Maintain and update the model after handover.

Chapter 18:

A Practical Framework for Delivering LOD 500

18.1 The Framework Overview

- **Phase 1: Define** (Pre-Design)
- **Phase 2: Plan** (Design)
- **Phase 3: Execute** (Construction)
- **Phase 4: Verify** (Throughout)
- **Phase 5: Handover** (Completion)
- **Phase 6: Sustain** (Post-Occupancy)

18.2 Phase 1: Define (Pre-Design)

- **Engage FM:** Identify FM team, understand needs, document requirements
- **Define Information Requirements:** What assets, what data, what format, what accuracy

- **Specify Standards:** ISO 19650, classification, naming, software, formats
- **Draft Contract Requirements:** Information deliverables, milestones, payment linkages, acceptance criteria
- **Budget for BIM:** Modelling, data collection, verification, software, training, support

18.3 Phase 2: Plan (Design)

- **Develop BIM Execution Plan:** One-page summary with clear responsibilities, timeline, and deliverables
- **Design for Information:** Include data fields, structure for FM, and plan for as-built updates.
- **Build Information Structure:** Create data templates, asset hierarchy, and test with sample data
- **Flow Requirements Down:** Include in subcontracts, specify for each trade, price BIM work
- **Prepare Systems:** Software, hardware, mobile tools, verification systems

18.4 Phase 3: Execute (Construction)

- **Integrate BIM with Site:** BIM staff on site, daily coordination, real-time updates
- **Capture Information Continuously:** Redline system, daily markups, weekly updates
- **Collect Data at Source:** At procurement, delivery, installation, commissioning
- **Coordinate Regularly:** Weekly meetings, model review, issue resolution, update assignments
- **Enforce Requirements:** Tie payment to information, withhold for non-compliance

18.5 Phase 4: Verify (Throughout)

- **Verify Progressively:** Monthly sample verification, compare model to site
- **Use Multiple Methods:** Visual inspection, laser scanning, photo documentation, data validation
- **Document Verification:** What was checked, when, by whom, results, corrections
- **Report on Compliance:** Monthly reports, progress against requirements, issues

18.6 Phase 5: Handover (Completion)

- **Prepare Handover Package:** Complete model, all data populated, verification evidence, manuals linked
- **Test with FM Systems:** Export to FM software, verify compatibility, test searches
- **Train FM Team:** Navigation, data access, updates, ongoing support
- **Conduct Joint Review:** FM reviews the model, sample verification together, and issues fixed
- **Deliver and Celebrate:** Formal handover, acknowledge team, document lessons

18.7 Phase 6: Sustain (Post-Occupancy)

- **Provide Support:** Help desk, error correction, and update assistance
- **Maintain the Model:** FM updates for changes, process for recording modifications
- **Capture Lessons:** What worked, what didn't, share with the next project.

18.8 Framework Summary

Phase	Focus	Key Activities
Define	What's needed	Engage FM, define requirements, specify standards, draft contracts, budget.
Plan	How to deliver	Develop BEP, information design, build structure, flow down, prepare systems.
Execute	Do the work	Integrate with the site, capture continuously, collect data, coordinate, enforce.
Verify	Check accuracy	Verify progressively, use multiple methods, document, report.
Handover	Transfer ownership	Prepare package, test with FM, train, review together, deliver.
Sustain	Maintain value	Provide support, maintain model, capture lessons.

Conclusion

The Choice We Face

This book has documented why BIM fails to deliver LOD 500. Every player, every process, every failure point. Contracts fail. Workflows break. Information dies.

The picture is bleak. But it doesn't have to be.

Every failure documented is fixable. Contracts can be rewritten. Workflows can be integrated. Information can be captured. People can be trained. Models can be verified. LOD 500 can be delivered.

The industry can continue as it is:

- Clients who don't know what they need
- Consultants who design in isolation
- Contractors who build without recording
- Subcontractors who leave with information
- BIM Managers without power
- FM teams that receive useless models

Or the industry can change:

- Clients who define requirements clearly
- Consultants who design for information
- Contractors who integrate BIM with construction
- Subcontractors who capture what they build
- BIM Managers with authority and resources
- FM teams that receive usable models

The choice is ours.

LOD 500 is possible. It is being delivered on some projects today. It could be delivered on yours.

The question is not, "Can we do it?" but, "Will we do it?"

About the Author

Abdul Rafiq Ganasan Bin Abdullah

Born and raised in Johor, Malaysia, Abdul Rafiq Ganasan bin Abdullah has spent thirty-five years navigating the complex world of architecture, engineering, and construction. His journey through the industry has given him a perspective that few possess—one forged not in academic theory alone but in the daily reality of projects that succeeded, failed, and everything in between.

Education and Professional Development

Rafiq's formal education began with a Diploma in Architecture, providing him with the foundational understanding of design, spatial relationships, and the built environment. Recognising early that the industry was changing, he pursued further specialisation in Building Information Modelling, earning a Professional Certificate in BIM.

His commitment to mastering the digital transformation of construction led him to my BIM Malaysia under CIDB (Construction Industry

Development Board), where he completed the rigorous pathway to become a Certified BIM Manager. This certification validated not just his technical knowledge, but his understanding of the processes, standards, and management practices required to lead BIM implementation on complex projects.

Today, he is an active member of the Malaysian Board of BIM (Pertubuhan BIM Malaysia), holding membership number P-00137. This affiliation connects him to the broader community of BIM professionals working to advance the practice in Malaysia and beyond.

Thirty-Five Years in the AEC Industry

For over three decades, Rafiq has worked across the full spectrum of the Architecture, Engineering, and Construction industry. He has seen projects from conception to completion, from the drawing board to the site, from handover to operation. This longevity has given him something rare: perspective.

He has witnessed the industry evolve from manual drafting to digital modelling. He has seen the promises of new technologies and the disappointments when those promises went unfulfilled. Most importantly, he has observed the patterns, the recurring issues that plague project after project, regardless of scale, budget, or team.

The Issues That Never Go Away

Through decades of practice, Rafiq has watched the industry cycle through the same problems:

- Delays that cascade through project schedules
- Disputes that burn time, money, and relationships
- Defects that appear despite quality processes
- Reworks that consume profit and morale
- Miscommunications that send teams in the wrong directions
- Discrepancies between what was designed and what was built

These issues are not new. They are not unique to any firm, country, or project type. They are systemic, and for decades, the industry has accepted them as inevitable.

The Digital Solution

Rafiq's conviction, born of thirty-five years of experience, is that these issues are not inevitable. They are solvable. The solution lies in the proper implementation of BIM processes, not as a software exercise, but as a fundamental transformation of how projects are delivered.

But BIM alone is not enough. Rafiq has seen too many projects invest in software, hire BIM staff and still fail to deliver. The missing element is full commitment and genuine understanding from every player in the project ecosystem.

When clients understand what they need. When consultants design for information, not just geometry. When contractors integrate BIM with site operations. When subcontractors capture what they build. When BIM Managers have real authority. When FM teams receive models they can actually use.

When all of this happens, the old issues—delays, disputes, defects, reworks, miscommunications, discrepancies—can be resolved digitally. Not eliminated, perhaps, but reduced to a fraction of their current impact.

Why This Book

This book is the product of thirty-five years of watching, learning, and concluding. It is not an academic treatise written from an ivory tower. It is a practitioner's guide, written by someone who has been in the trenches, who has seen BIM succeed and fail, and who has drawn the lessons that only experience can teach.

Rafiq writes not to assign blame, but to assign responsibility. Not to complain about the industry, but to offer a way forward. Not to sell BIM as magic, but to show what it actually requires.

The audience for this book is everyone in the AEC industry: clients, consultants, contractors, subcontractors, BIM Managers, and facility managers. Each chapter speaks directly to one of these players, naming the ways they contribute to failure and offering a path to success.

A Personal Note

In his thirty-five years, Rafiq has worked alongside countless professionals—some brilliant, some struggling, all doing their best in a demanding industry. He has seen the frustration on the face of a BIM Manager whose warnings were ignored. He has heard the resignation in the voice of a Facility Manager handed a useless model. He has felt the tension in meetings where disputes could have been avoided with better information.

This book is written for them, for the people on the ground who deal with the consequences of failed BIM delivery every day, and for the leaders who have the power to change how things are done.

The industry can do better. Rafiq believes this deeply. This book is his contribution to making it happen.

You can connect with Abdul Rafiq Ganasan bin Abdullah through the Malaysian Board of BIM (Pertubuhan BIM Malaysia), where he continues to advocate for better BIM implementation across the Malaysian construction industry and beyond.

Member, Malaysian Board of BIM (P-00137)

Certified BIM Manager, myBIM Malaysia / CIDB

Johor, Malaysia

rafiqganasan1604@gmail.com

www.ingramcontent.com/pod-product-compliance
Lightning Source LLC
LaVergne TN
LVHW050923210926
848482LV00012B/512
* 9 7 9 8 9 0 3 4 2 6 3 4 8 *